PRIMARY GEOBOARD
ACTIVITY BOOK

(Grades K–3)

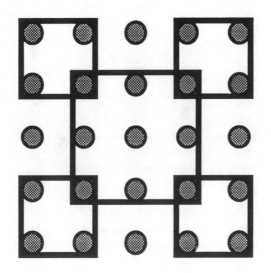

⚠ **WARNING:**
CHOKING HAZARD - Small parts.
Not for children under 3 years.

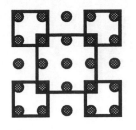 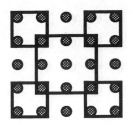

Introduction

The *Primary Geoboard Activity Book* is a resource book of reproducible blackline masters designed to help K–3 teachers make the best use of the geoboard. Children (and teachers, too!) are intrigued by the colorful rubber bands and the geometric symmetry of the geoboard.

In this book, teachers in the primary grades will find activities that go far beyond using the geoboard to teach the basic ideas of geometry. Exploring in a directed setting or individually, children will begin to learn the key concepts for area, perimeter, fractions, congruence, and elementary transformational geometry.

Each of the seven sections in the *Primary Geoboard Activity Book* focuses on a related group of activities. The blackline masters build sequentially on students' prior learning and involve copying, manipulating or designing on the geoboard. As students learn geometry via the manipulation, their visual thinking and spatial reasoning skills will improve. Students will also refine problem-solving strategies, particularly in terms of guessing, checking and revising their work.

Teaching suggestions are outlined at the beginning of each section:

Getting Started	Pre-learning activities to model concepts and show why they are important.
Using the Worksheets	Pacing and additional teaching hints.
Practice	Extension activities for cooperative learning to put the concepts into practice.
Wrap-up	Checks for understanding and use of terms.

Younger children using geoboards for the first time may have trouble associating the geoboard pegs with the corresponding dots on the worksheets. To help those having difficulty, put your finger on one dot at a time and have students point to the corresponding peg.

Table of Contents

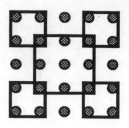

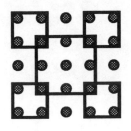

Table of Contents

Teacher's Notes

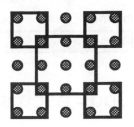

Section A
Shapes, Corners, Sides

Getting Started   Use a geoboard to show children various-sized squares and rectangles. Ask them to count the corners and sides with you. Help the children to see that the number of corners equals the number of sides. Ask children why they think this is so. Children may say that a side is added each time the geoband is wrapped around a new peg or corner.

Using the Worksheets

Use the worksheets in the order shown at the right.	Copying Squares and Rectangles	2
	Copying Triangles	3
	Copying Shapes	4
	Changing Shapes	5
	Making More Corners	6
	Making Your Own Shapes	7
	Counting Sides	8

As you use the worksheets, begin each activity by having children copy shapes that you make. Children should not be concerned about matching the locations of your shapes. Instead, emphasize the same number of corners and sides, as well as the general shape of each figure.

Eventually children will learn to match the geoboard pegs with the corresponding dots on the worksheets. For a child having trouble orienting, put your finger on one dot at a time and have the child point to the corresponding peg.

Practice   Have the children make squares and rectangles of red, green, yellow, and blue geobands. They can count the pegs aloud as they attach the geobands. As you progress to triangles and irregular shapes, the children can work in pairs, taking turns making new shapes with different-colored geobands.

Wrap-up   Ask children to make a specific figure on their geoboards. For example, tell them to make a red shape that touches only 3 pegs, and so forth. Ask: How many corners? How many sides?

Copy each square or rectangle on your geoboard.

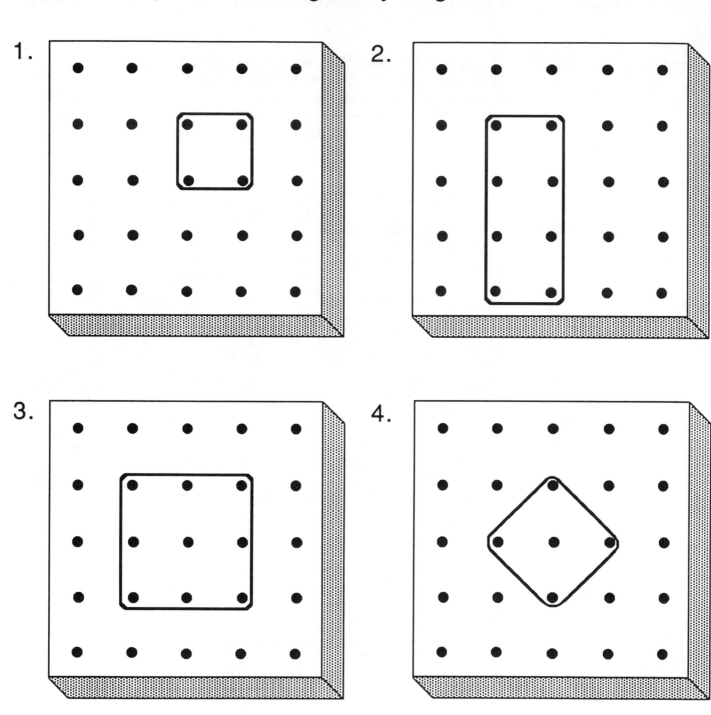

1.

2.

3.

4.

5. Which shape do you like the best? _____

2

Copy each triangle.

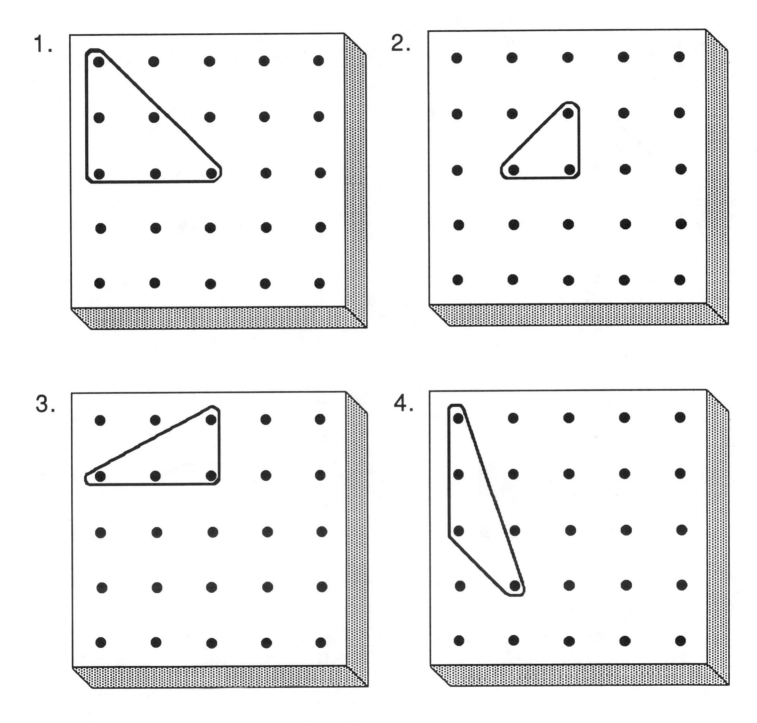

1.

2.

3.

4.

5. Which triangle is the smallest? _____

Copy each shape.

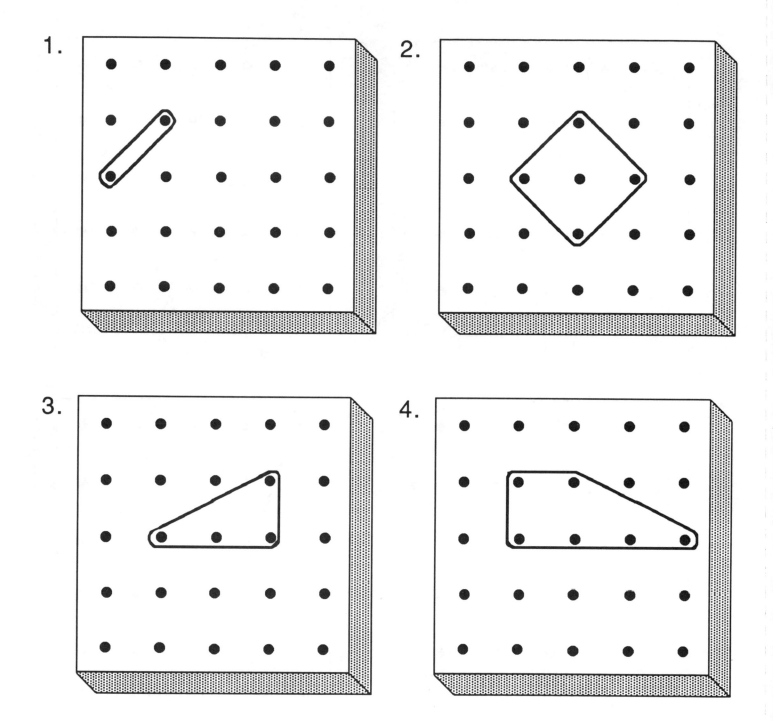

1.

2.

3.

4.

Color your favorite shape blue.

Change Shape A into Shape B.

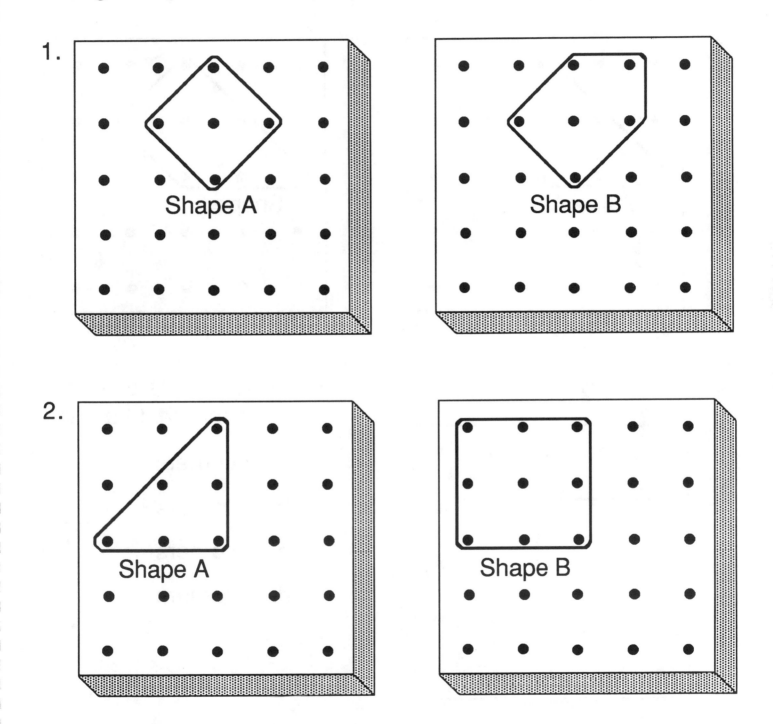

1.

Shape A

Shape B

2.

Shape A

Shape B

Color Shape A red. Color Shape B green.

Change Shape A into Shape B.

1.

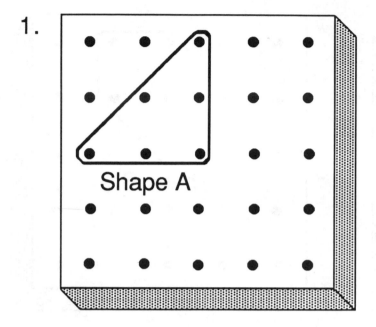

Shape A

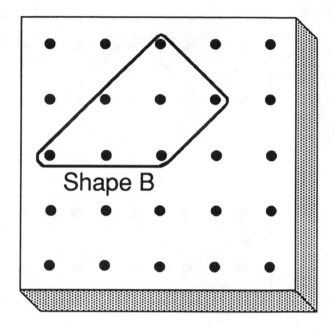

Shape B

2. How many corners? _____

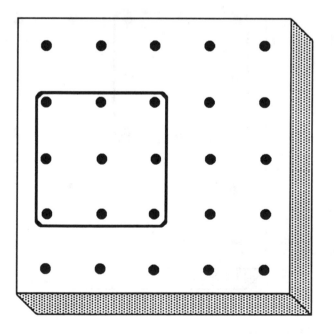

3. Make shapes on your geoboard that have:

a. 3 corners

b. 4 corners

c. 5 corners

d. 6 corners

Make and draw shapes that touch only 3 pegs.

1.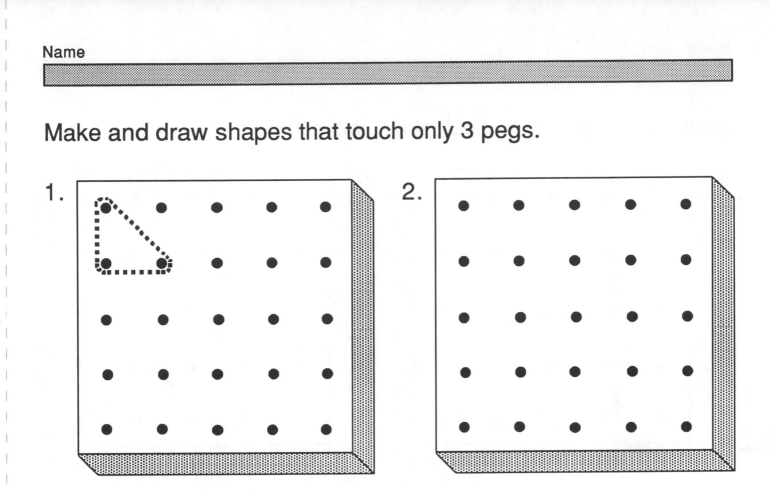

2.

Make and draw shapes that touch only 4 pegs.

3.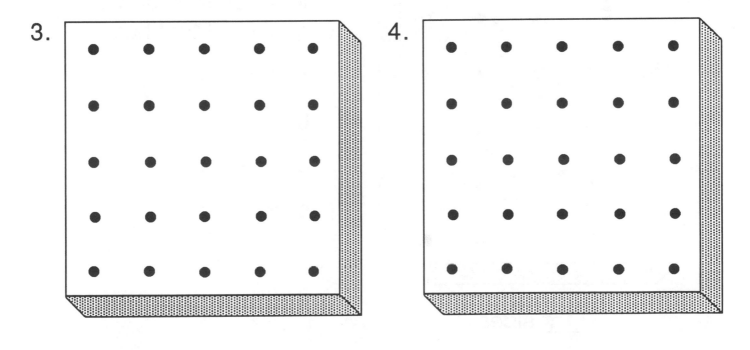

4.

How many sides does each shape have?

1.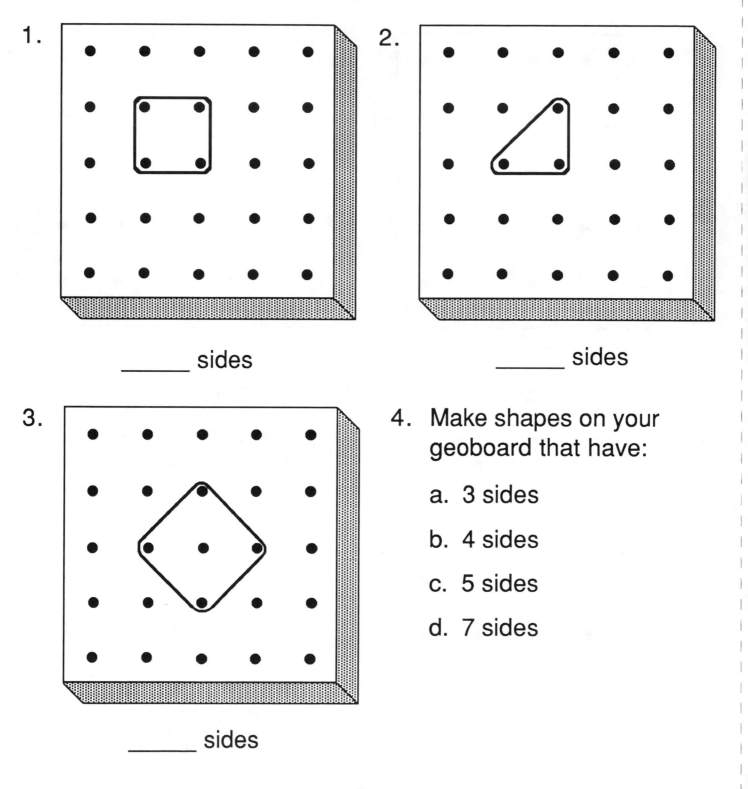

_____ sides

2.

_____ sides

3.

_____ sides

4. Make shapes on your geoboard that have:

a. 3 sides

b. 4 sides

c. 5 sides

d. 7 sides

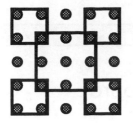

Teacher's Notes

Section B
Shrinks and Stretches

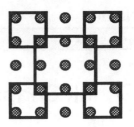

Getting Started

Remind children how a geoband can stretch. Then use your geoboard to stretch a small square into a larger one. Point out that the number of corners and sides remains the same as you stretch or shrink a figure.

Using the Worksheets

	Shrinking Figures	10
	Shrinking More Figures	11
Use the worksheets	Shrinking Triangles	12
in the order shown	Stretching Figures	13
at the right.	Stretching More Figures	14
	Double-Stretching Figures	15
	Shrunk or Stretched?	16

As you progress through the worksheets, have children create the stretches or shrinks by moving the fewest corners of the geoband. Have them count the pegs enclosed by the geoband each time.

Have children describe the figures they make. Have them tell what is alike and different about the original shape and the stretched or shrunken shape (i.e., longer or shorter sides). Children should also identify figures that are too small to be shrunk any more.

Practice

Have children work in pairs to shrink and stretch figures. Have one child design a figure with three or four sides. The designer may decide whether the partner should shrink or stretch the figure. Both children can compare the number of pegs needed for the two shapes.

Wrap-up

Show children figures that you then stretch or shrink, and have them tell which you did. Then show children pairs of figures on your geoboard labeled A and B. Ask whether figure B is stretched or shrunken from figure A, and vice-versa. Then ask the children to shrink or stretch the figures you show on your geoboard.

Shrink Shape A to the size of Shape B on your geoboard.

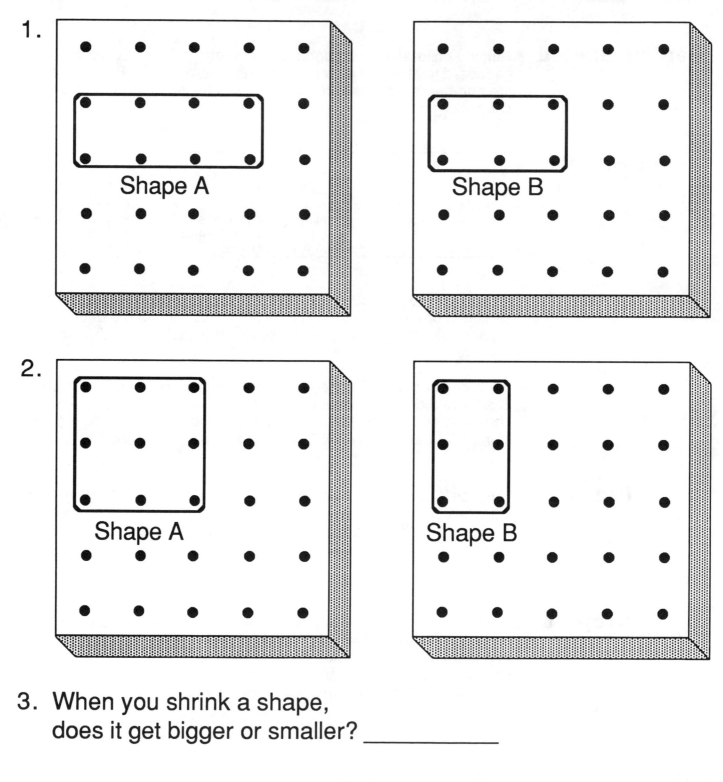

1. Shape A   Shape B

2. Shape A   Shape B

3. When you shrink a shape,
   does it get bigger or smaller? _____

Shrink Shape A to the size of Shape B.

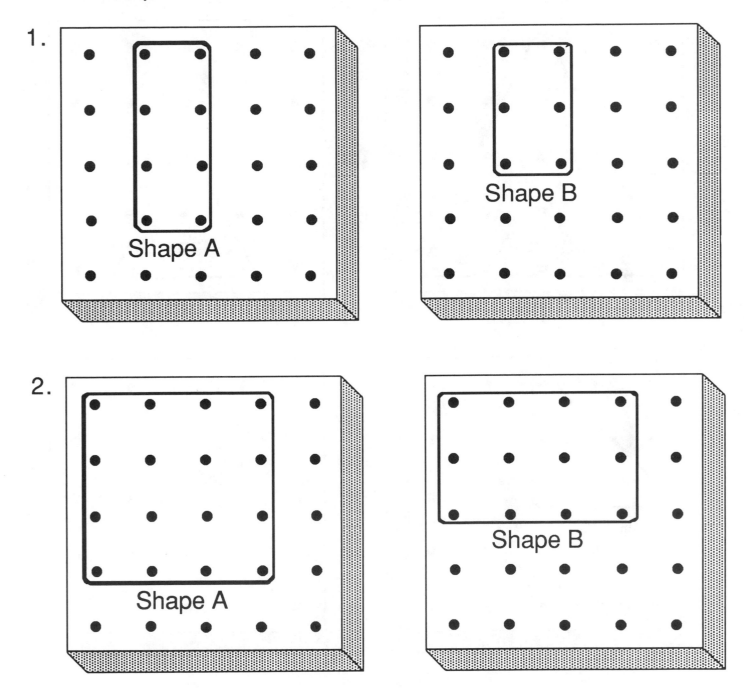

1.

Shape A

Shape B

2.

Shape A

Shape B

Now try to shrink Shape B even smaller on your geoboard.

Shrink the large shape to match the small shape.

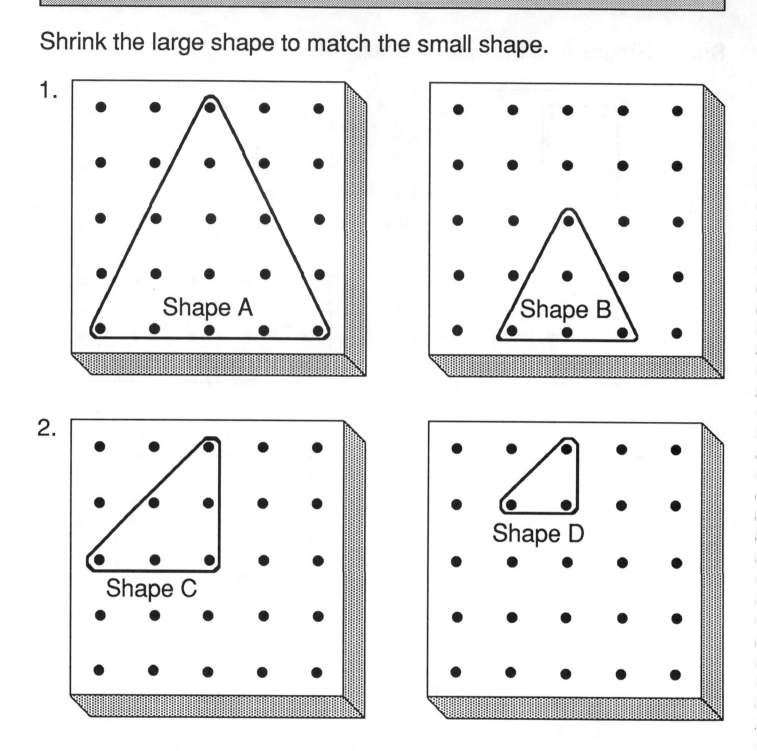

1.

Shape A

Shape B

2.

Shape C

Shape D

3. Which two shapes have a square corner? _____ _____

Stretch Shape A to the size of Shape B.

1.

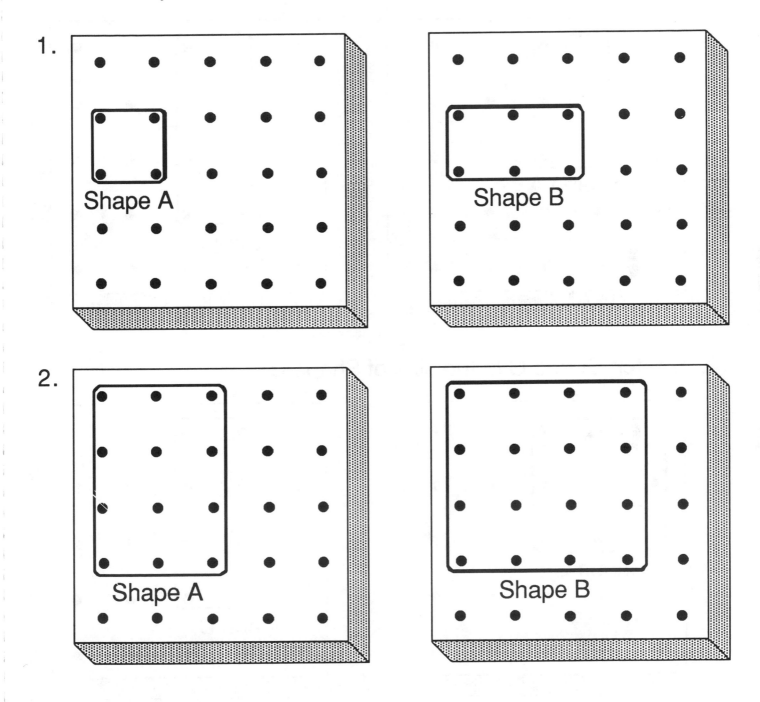

Shape A

Shape B

2.

Shape A

Shape B

3. When you stretch a shape
   does it get bigger or smaller? _____

1. Stretch Shape A to the size of Shape B.

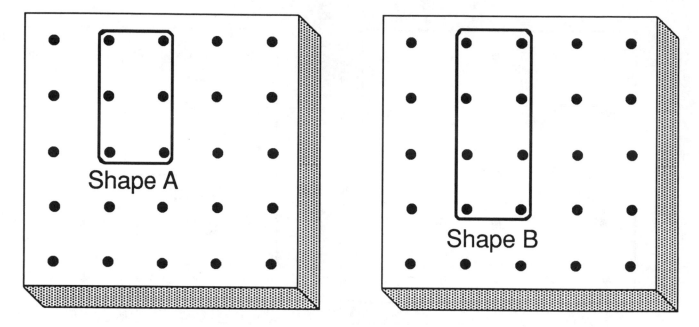

Shape A

Shape B

2. Stretch Shape C to the size of Shape D.

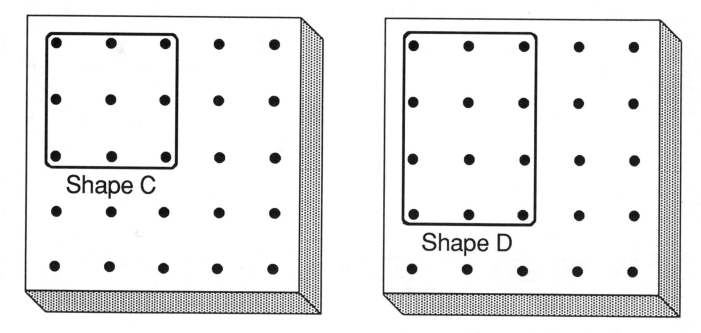

Shape C

Shape D

3. Now stretch Shape A to the size of Shape D.

Stretch Shape A to the size of Shape B.

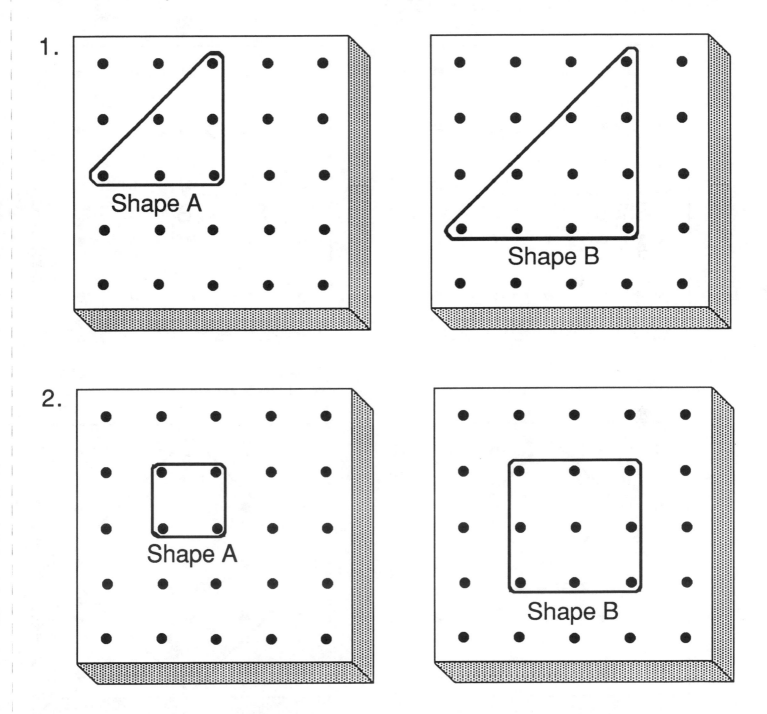

1.

Shape A

Shape B

2.

Shape A

Shape B

Now try to stretch Shape B even larger on your geoboard.

Copy Shape A. Now copy Shape B. Is Shape B shrunk or stretched? Circle the answer.

1.

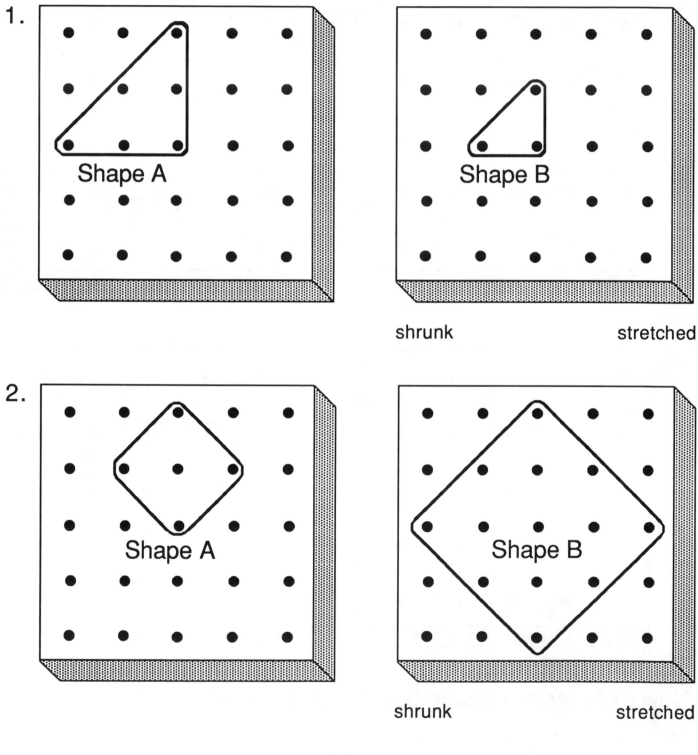

Shape A

Shape B

shrunk   stretched

2.

Shape A

Shape B

shrunk   stretched

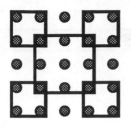 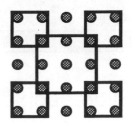

Teacher's Notes

Section C
Slides and Congruence

Getting Started

Use desks or chairs set in rows and columns to mimic the five-by-five geoboard peg configuration as closely as possible. Have children stand beside the "pegs" at various places. Have them slide one space left, right, up, or down at your direction. First have children do this alone, taking turns one at a time; then in pairs; then all together. Show children their moves by tracing them with your finger on a geoboard.

Using the Worksheets

Use the worksheets in the order shown at the right.	Sliding Right	18
	Sliding Left	19
	Sliding Down	20
	Sliding Up	21
	Sliding Two Spaces	22
	How Many Spaces?	23
	Did the Same Shape Slide?	24

As you progress through the worksheets, check to see that children do not stretch or shrink the shapes as they slide them. Have children describe each slide before and after they make it, such as *one space right* or *two spaces down.* Help children see that each figure is still the same shape and looks exactly like the original figure after sliding—only the location has changed.

Practice

Have children work in pairs to do slides. One partner will design the original figure and tell the direction and amount of the slide. The other partner shows the slide with a second, different-colored geoband, keeping the original figure in position.

Wrap-up

Show children a figure on your geoboard. First have them all do the same slide. Then have each child do a different slide and tell how it differs from the location of your original figure.

Slide each shape 1 space to the right.

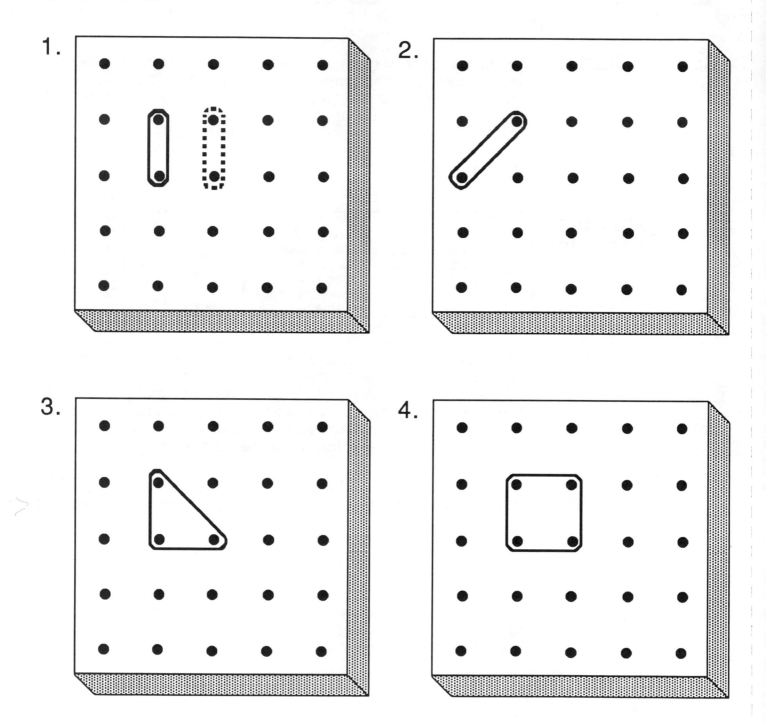

1.

2.

3.

4.

Slide each shape 1 space to the left.

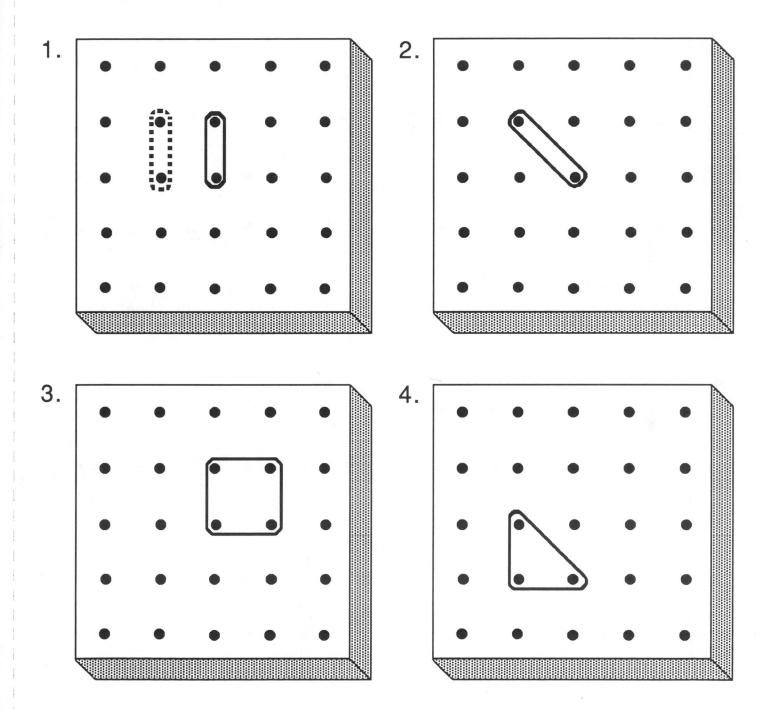

1.

2.

3.

4.

Shade each shape with your favorite color.

Slide each shape 1 space down.

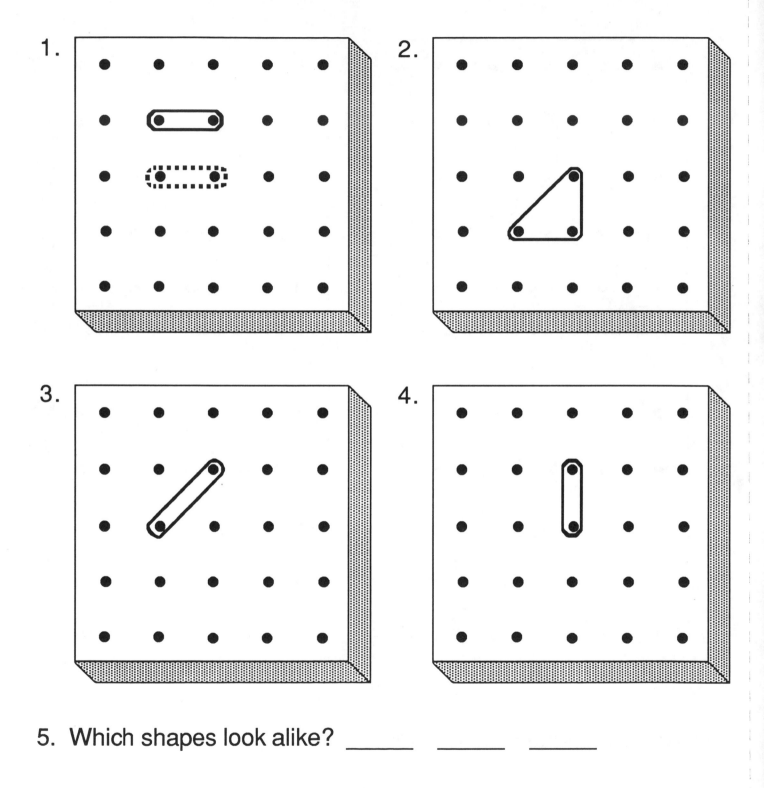

1.

2.

3.

4.

5. Which shapes look alike? _____ _____ _____

Slide each shape 1 space up.

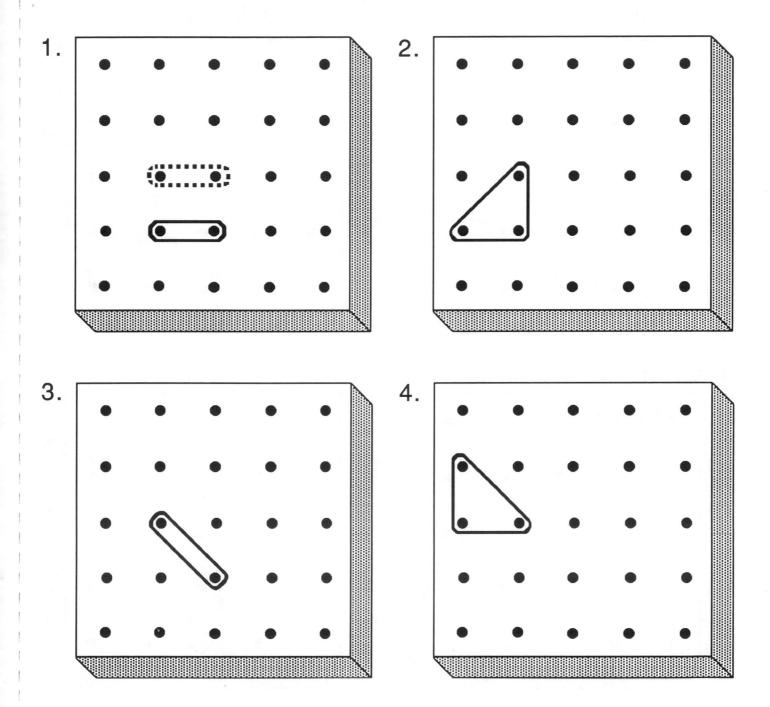

1.

2.

3.

4.

Try sliding the shapes in other ways on your geoboard.

Slide each shape 2 spaces.

1. To the right

2. To the left

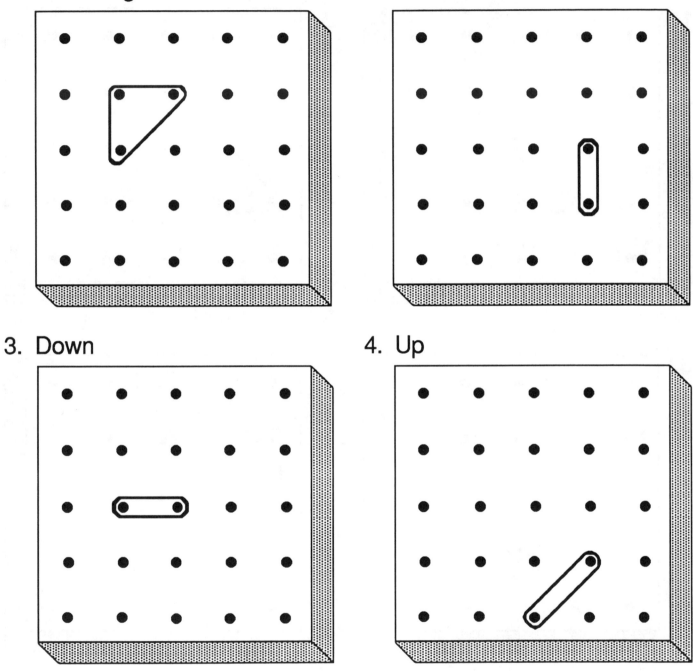

3. Down

4. Up

5. Which way is easiest for you to slide shapes? _____

Slide each shape. How many spaces did it slide?

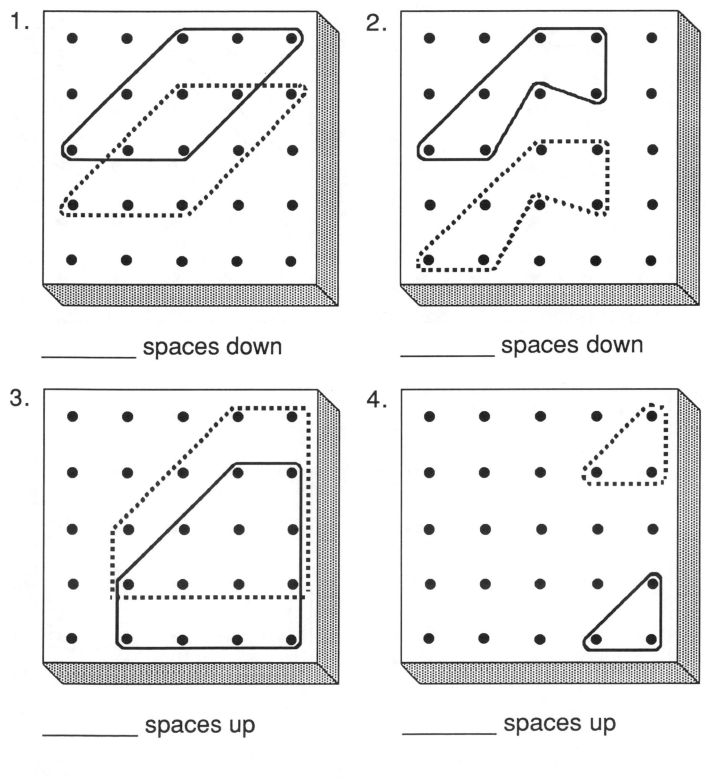

1.

_____ spaces down

2.

_____ spaces down

3.

_____ spaces up

4.

_____ spaces up

Are the shapes the same? Circle yes or no.

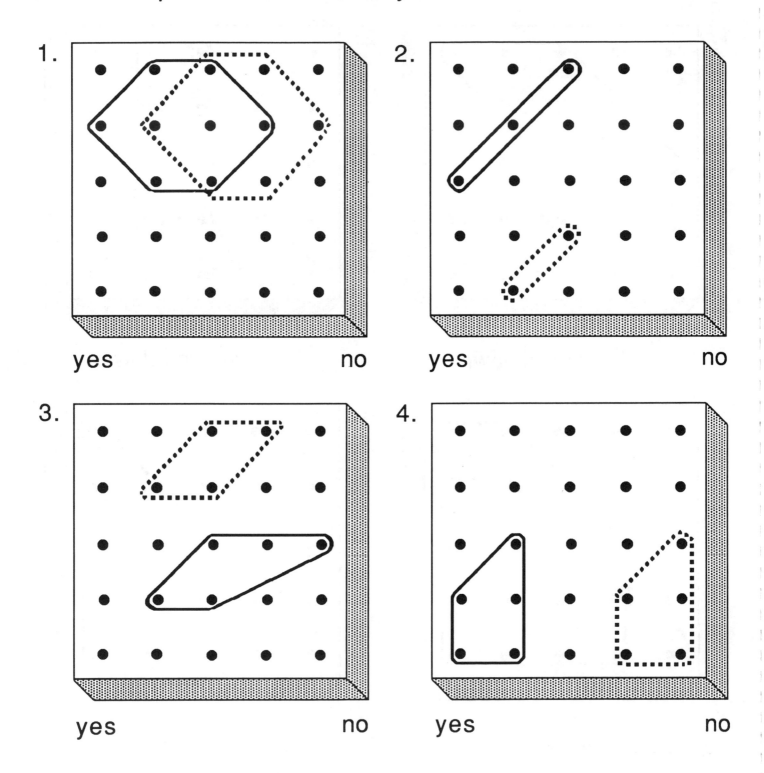

1.

yes   no

2.

yes   no

3.

yes   no

4.

yes   no

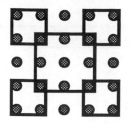

Teacher's Notes

Section D
Flips and Symmetry

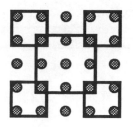

Getting Started

Hold your geoboard in front of you as you divide it in half by wrapping the middle row with a red geoband. Make a simple figure such as a line or square on the left, and then show children its twin on the right. Repeat until children volunteer to make the twin figures.

Then show children a symmetrical figure such as a rectangle on your geoboard. Use a red geoband to divide the figure in half. Show children how the red geoband makes two equal parts out of the figure. Continue with other symmetrical shapes until children volunteer to divide the figures into two equal parts.

Using the Worksheets

Use the worksheets in the order shown at the right.	Twin Flips	26
	More Twin Flips	27
	Flips Over a Line	28
	Flips of Bigger Shapes	29
	Mirror Parts	30
	Two Equal Parts?	31
	Equal Parts Two Ways	32

As you progress through the worksheets, first provide the initial figure for children to mirror. Then have children copy the original figure and its twin, and finally make the twin on their own. For Worksheets 31 and 32, some figures will have more than one correct line of symmetry. It is acceptable for children to see and indicate only one line of symmetry in each figure.

Practice

Have children plan twin figures of their own. One child makes a figure and passes it to another child who makes the twin. A third child attaches a red geoband as the divider. Do a similar activity with symmetrical figures.

Wrap-up

Ask children to make the twins of figures you make. Then make symmetrical figures for them to divide into two equal parts.

Copy each shape. Then make its twin.

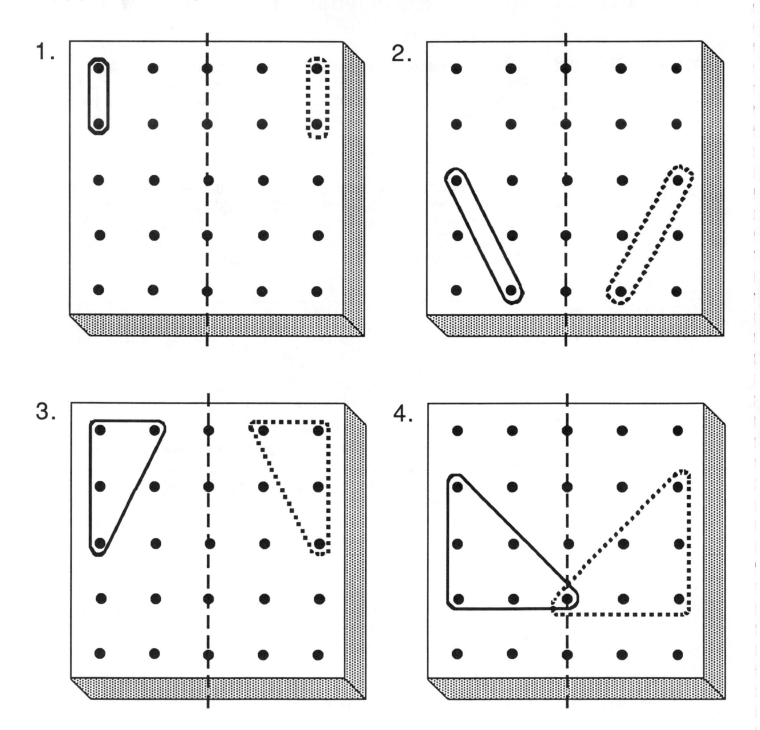

1.

2.

3.

4.

Copy each shape. Then make and draw its twin.

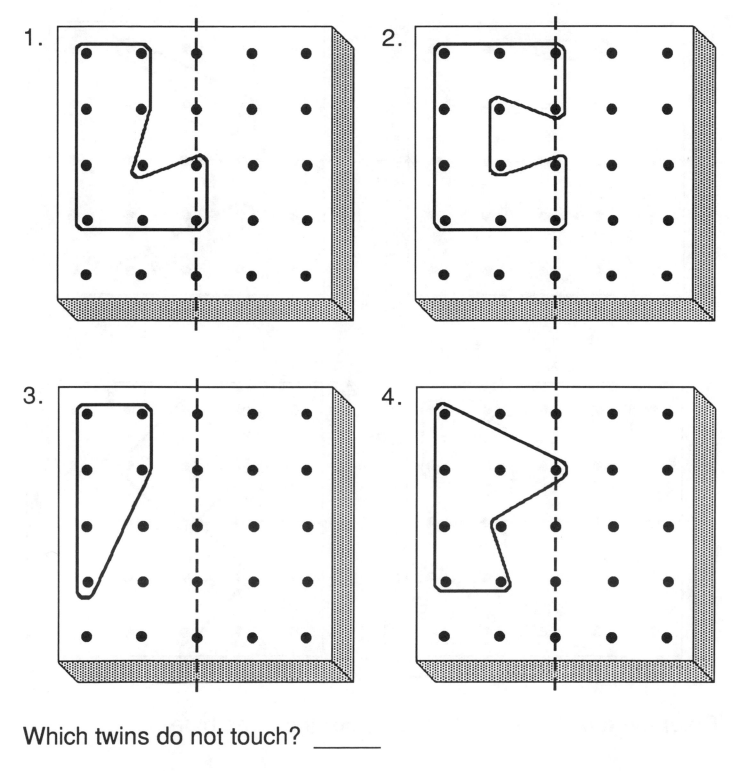

1.

2.

3.

4.

Which twins do not touch? _____

Copy each shape. Then make its twin.

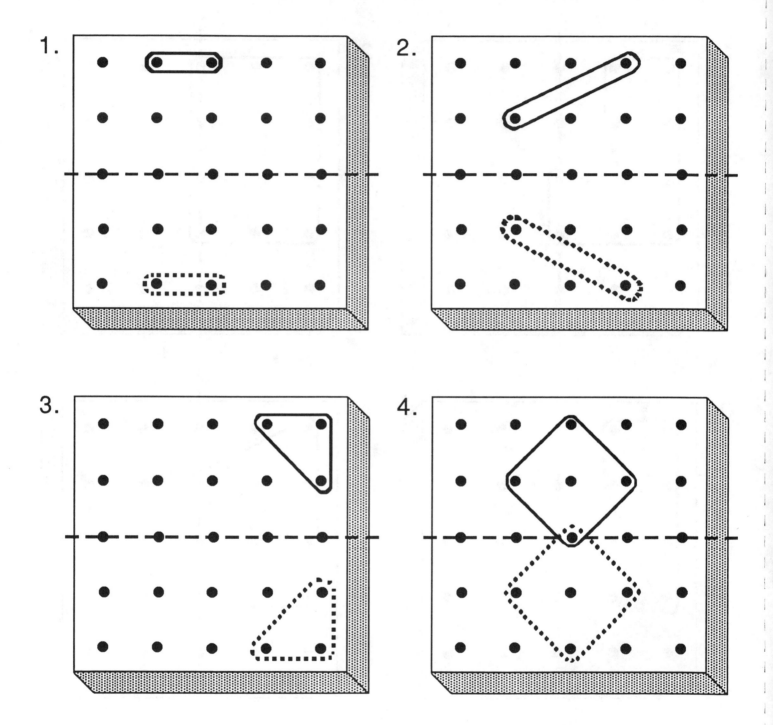

Color the top shapes red. Then color the twins blue.

Copy each shape. Then make and draw its twin.

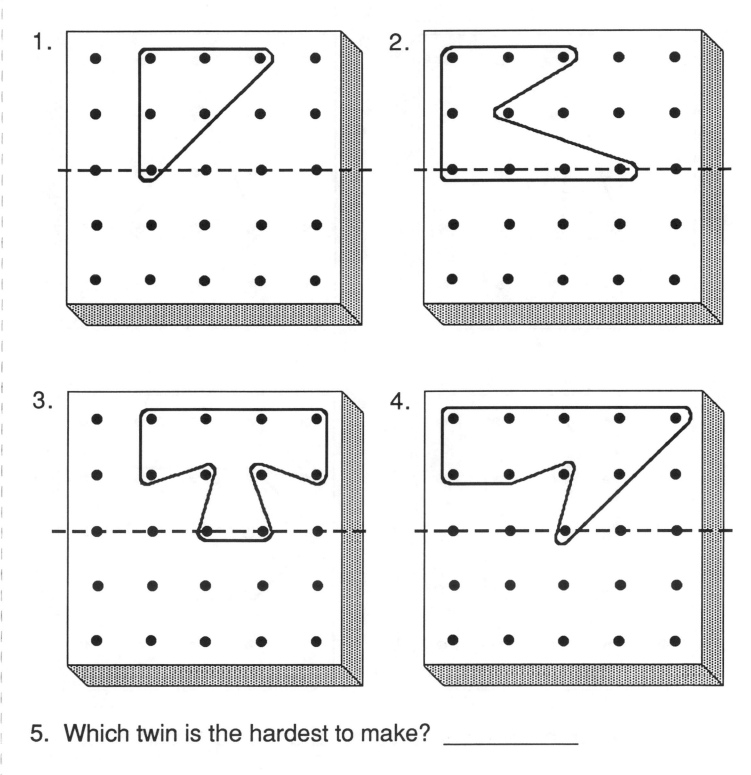

1.

2.

3.

4.

5. Which twin is the hardest to make? _____

Copy each shape.
Then use a geoband to make two equal parts.

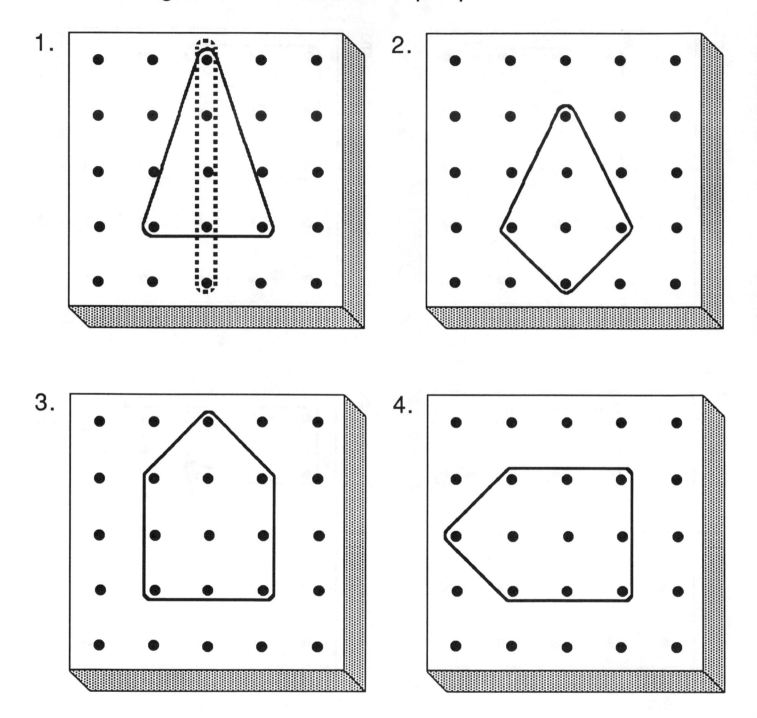

1.

2.

3.

4.

Shade one part of each shape yellow and the other part blue.

Copy each shape. Can you make two equal parts?
Circle yes or no.

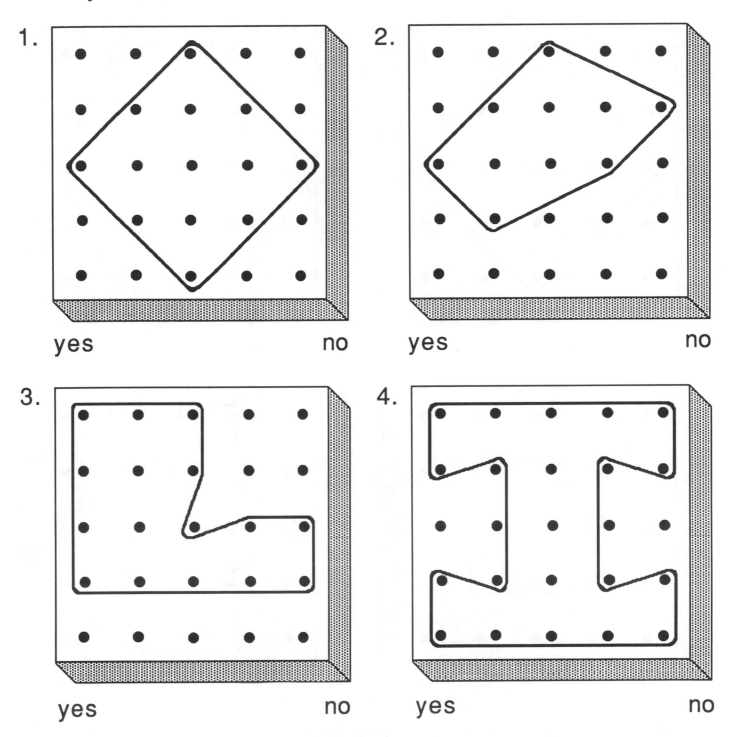

1.

yes   no

2.

yes   no

3.

yes   no

4.

yes   no

Use two geobands of different colors.
Make equal parts two ways.

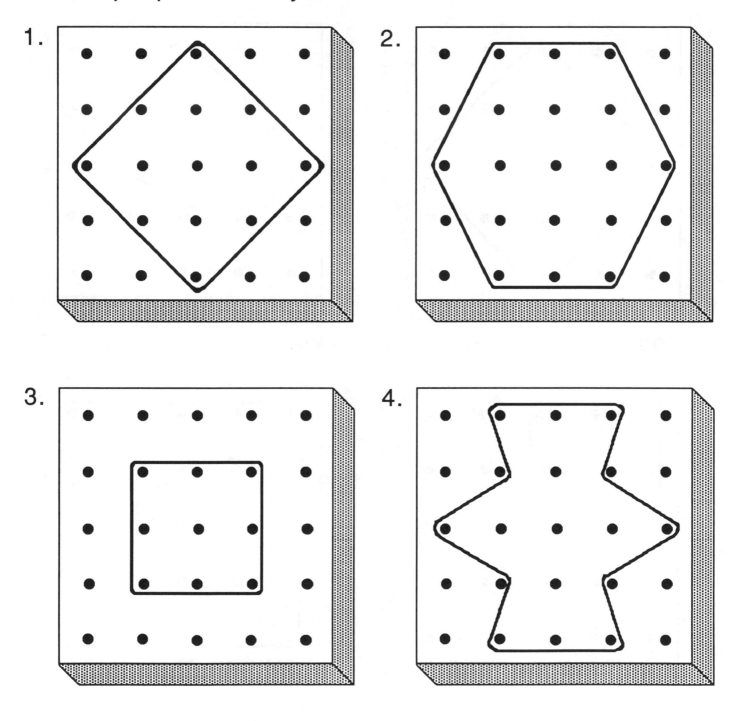

1.

2.

3.

4.

5. Which shape is the the smallest? _____

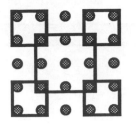

Teacher's Notes

Section E
Area

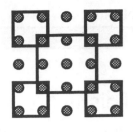

Getting Started

Use your geoboard and a geoband to mark off a one-unit square. Show children the raised lines on their geoboards that mark off unit squares. Have them trace the lines with their fingers. You may wish to show children a construction-paper square that is the same size as a unit square on the geoboard. Then make several figures on your geoboard and help children count the unit squares in each figure.

Using the Worksheets

Use the worksheets in the order shown at the right.	Unit Squares	34
	More Shapes with Unit Squares	35
	Counting Unit Squares	36
	Using Six Unit Squares	37
	How Many Unit Squares?	38
	How Big is Each Shape?	39
	Making Shapes of Different Sizes	40

As you progress through the worksheets, make sure children are counting the unit squares correctly, not missing any or counting any twice. Have them trace and cut out some shapes and mark off the unit squares on the cutouts. Introduce the word *area,* and explain that it means the number of square units that cover a shape.

Practice

Have children make figures with a specific number of unit squares. Have the class see how many different figures are made from 3, 5, 7, and 10 unit squares. Some children may want to make multiple figures of fewer unit squares to reach a total such as 10.

Wrap-up

Make a figure on your geoboard. Ask children to first estimate, then count the number of unit squares in it. Then have children make the same figure on their geoboards. Finally, have them each try to make a different figure using the same number of unit squares.

This is 1 unit square.
Make shapes with 2 unit squares.

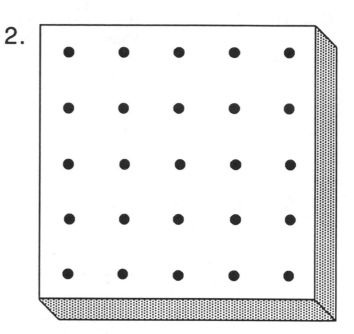

1.
2.

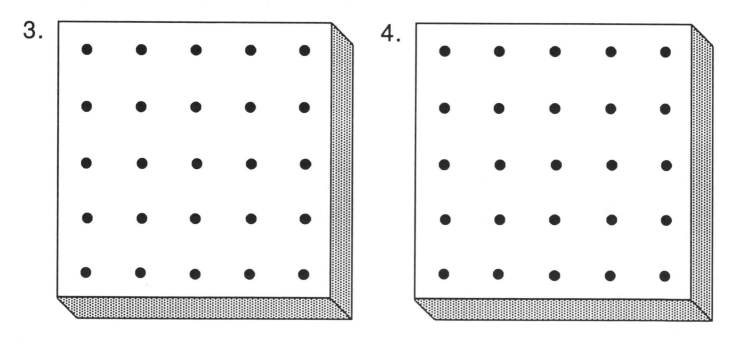

Make shapes with 3 unit squares.

3.
4.

Make shapes with 4 unit squares.

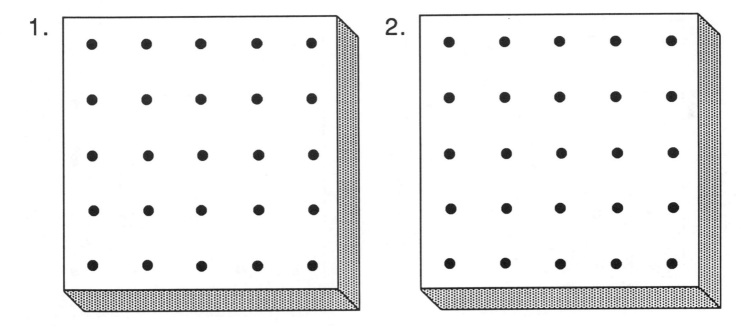

1.

2.

Make shapes with 5 unit squares.

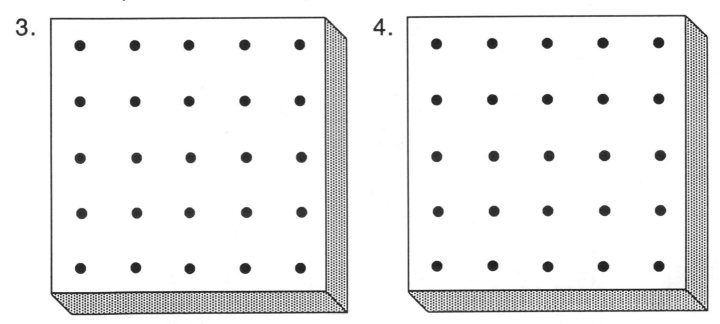

3.

4.

Shade each shape a different color.

How many unit squares?

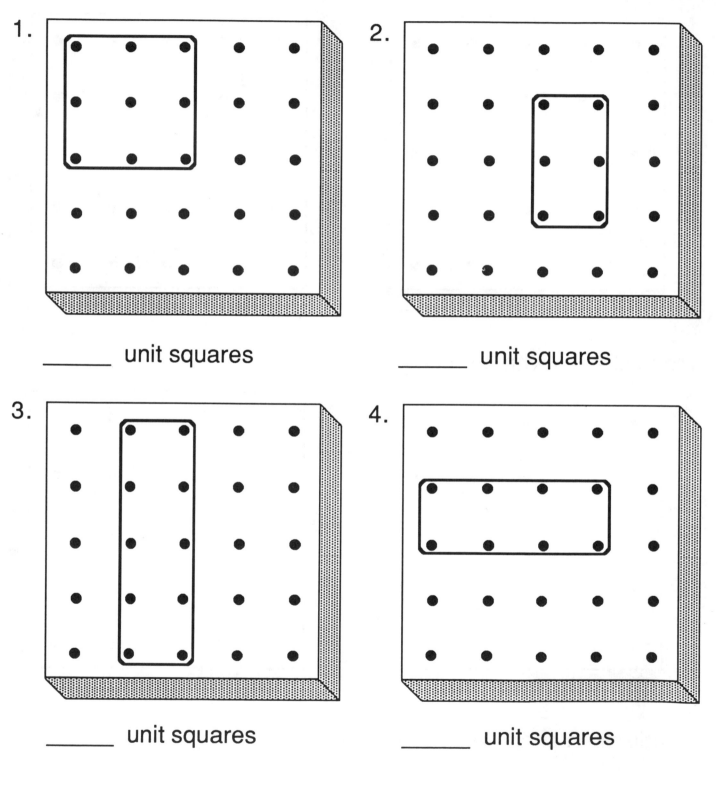

1. _____ unit squares

2. _____ unit squares

3. _____ unit squares

4. _____ unit squares

Make shapes with 6 unit squares.

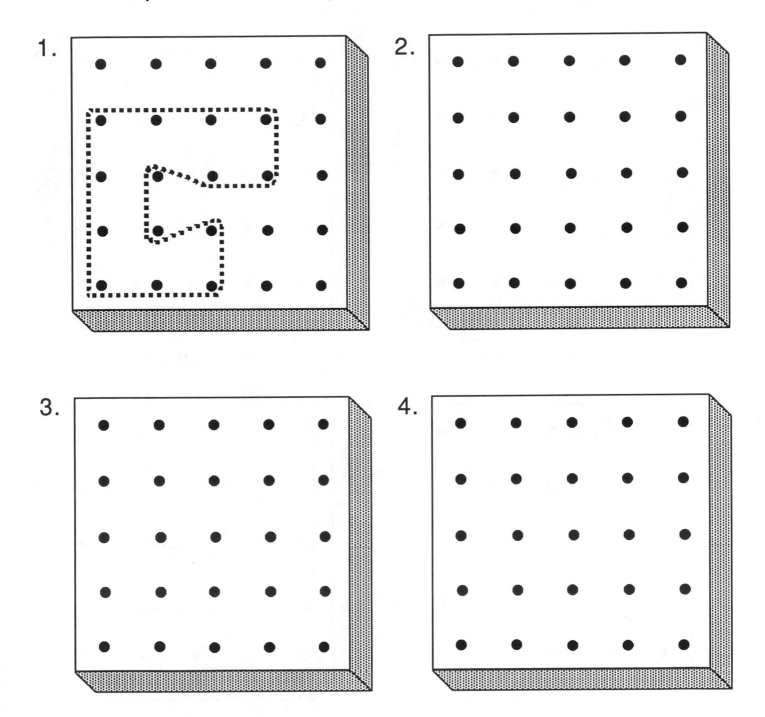

1.

2.

3.

4.

5. Which shape is your favorite? _____

Name

Copy each shape.
Count the unit squares to find each area.

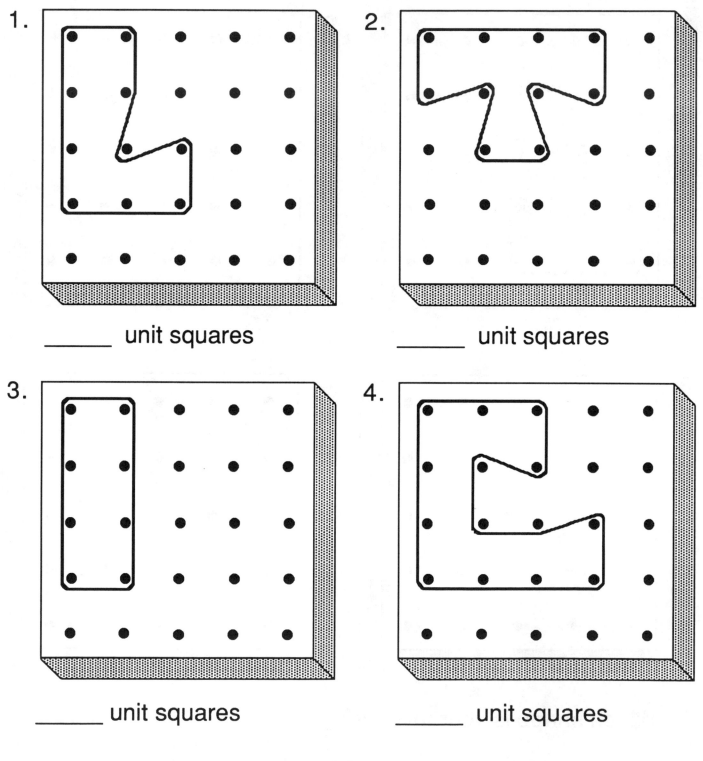

1. _____ unit squares

2. _____ unit squares

3. _____ unit squares

4. _____ unit squares

Name

Make the shapes.
Then match each shape with its area.

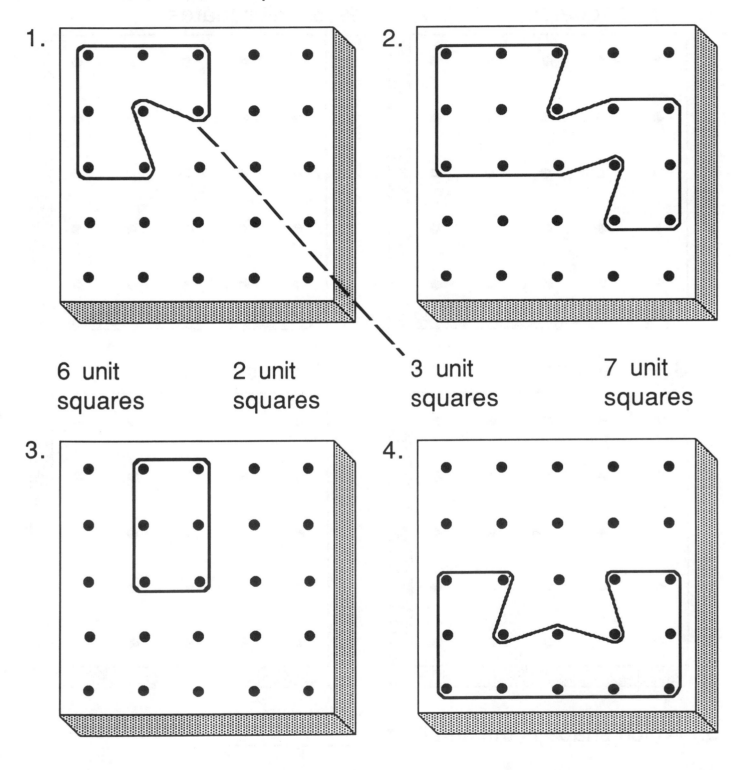

1.

2.

6 unit
squares

2 unit
squares

3 unit
squares

7 unit
squares

3.

4.

Make a shape with each area.

1. 4 unit squares

2. 5 unit squares

3. 3 unit squares

4. 7 unit squares

5. Which shape is the smallest? _____

Teacher's Notes

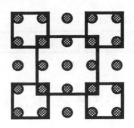

Section F
Fractions

Getting Started

Show children familiar, solid objects that you divide into halves and quarters, such as an apple, a piece of construction paper, and a slice of bread. Count 2 halves and 4 fourths. Have them tell how many people could share, each having one-half, and how many could each have one-fourth.

Then review dividing figures into two equal parts. Trace each part with your finger as you tell children that each part is *one-half.* Then use shapes drawn on the chalkboard. Have children take turns coming up and coloring each half with a different color. Then repeat each activity with fourths. Have children color one-fourth.

Using the Worksheets

Use the worksheets in the order shown at the right.	Halves	42
	More Halves	43
	Fourths	44
	Three-Fourths	45
	Halves or Fourths?	46
	Fractions for Shapes	47
	Equal Fractions	48

As you progress through the worksheets, help children realize that fractional parts can add up to a whole: two halves, four fourths. Point out that one-half of a figure is bigger than one-fourth of the same figure. Worksheet 48, introducing the important concept of equal fractions, may be challenging for some children. You may wish to use this page in a small-group setting.

Practice

Have children use yellow geobands to make figures that they think can be divided into equal halves and fourths. Have them use a red geoband to divide each figure into halves, and two blue geobands to divide the same figure into fourths.

Wrap-up

Show children figures you have divided into halves and fourths. Have them identify parts such as one-half, one-fourth, and three-fourths. Then provide figures which children can divide first into halves, then into fourths.

Make 2 equal parts.

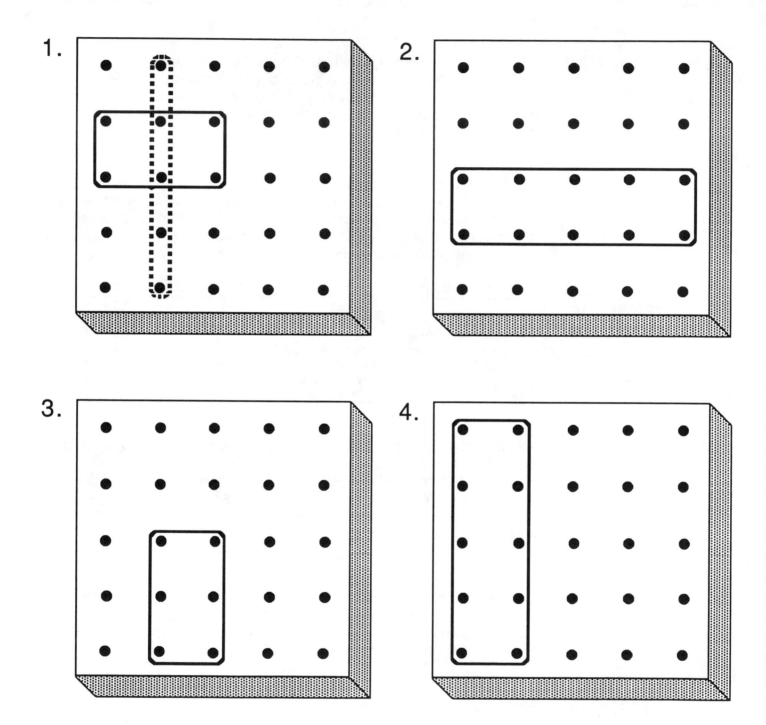

1.

2.

3.

4.

Color one-half of each shape blue.

Make 2 equal parts.

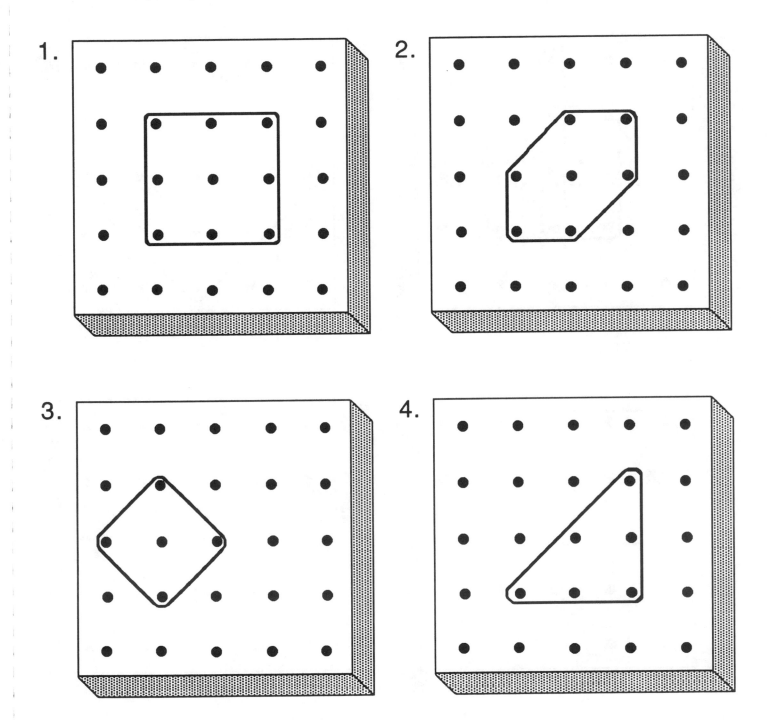

1.

2.

3.

4.

Color one-half of each shape red.

Make 4 equal parts.

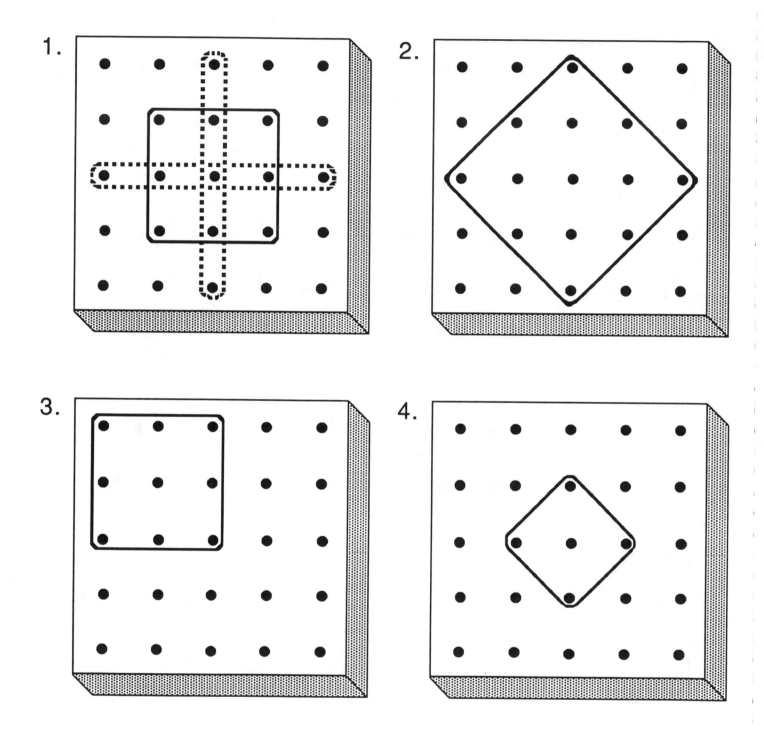

1.

2.

3.

4.

Color one-fourth of each shape green.

Make 4 equal parts.

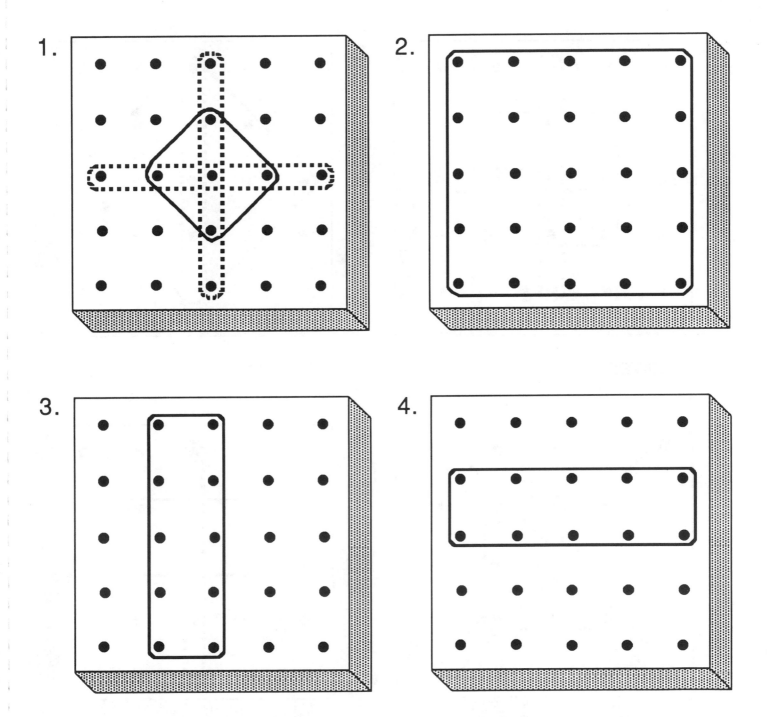

1.

2.

3.

4.

Color three-fourths of each shape yellow.

45

Copy the shapes on your geoboard. Circle halves or fourths.

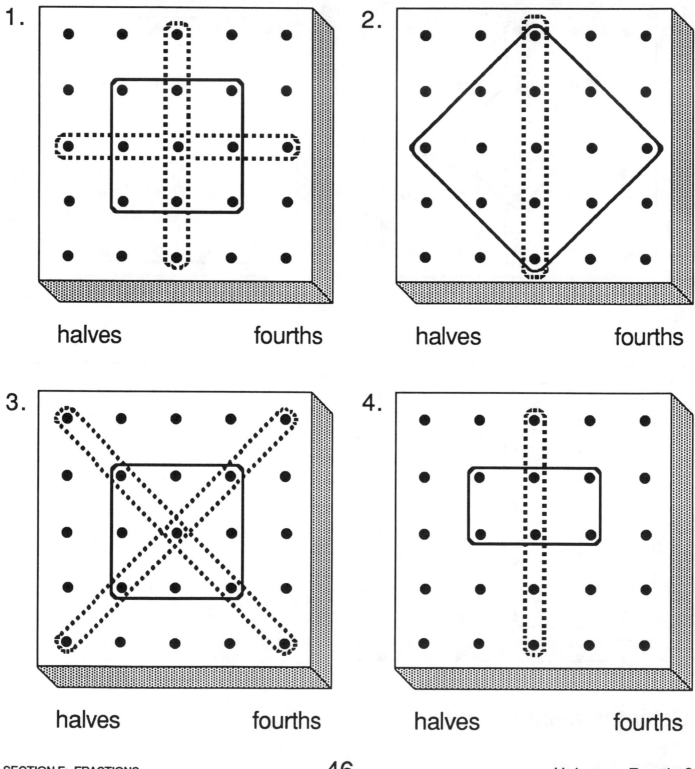

1. halves   fourths

2. halves   fourths

3. halves   fourths

4. halves   fourths

Copy the shapes. Then match the shaded area of each geoboard to a fraction.

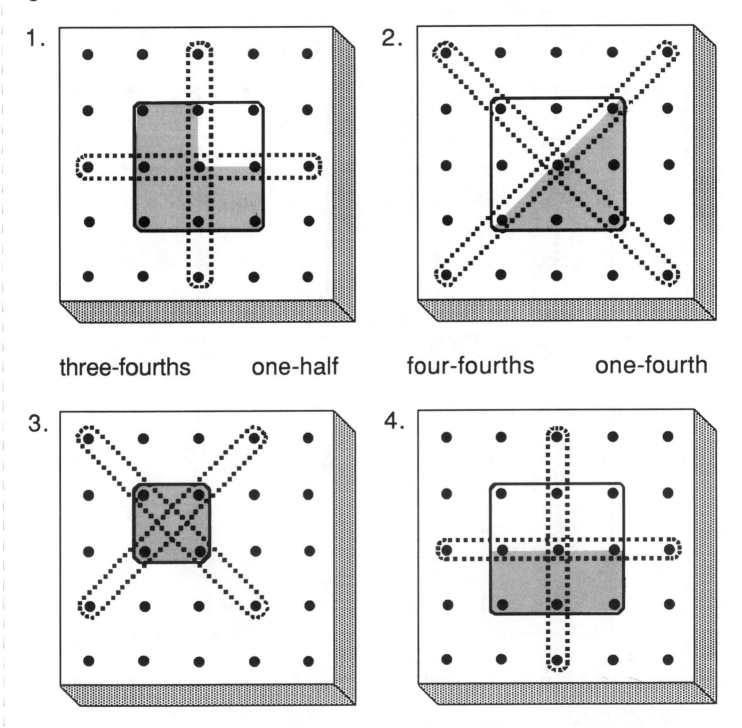

1.

2.

three-fourths   one-half   four-fourths   one-fourth

3.

4.

Name

Color to show each fraction.
Then match to show the equal fractions.

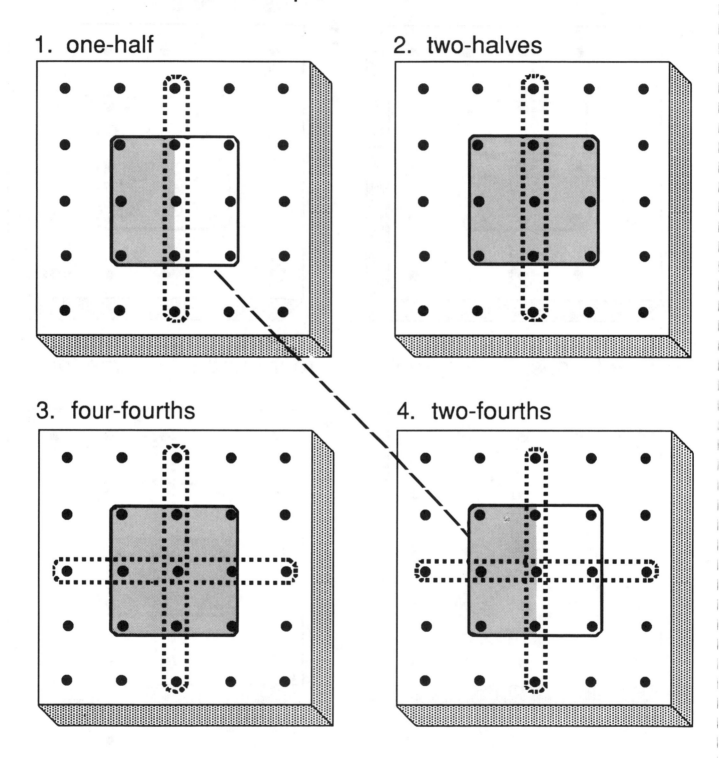

1. one-half

2. two-halves

3. four-fourths

4. two-fourths

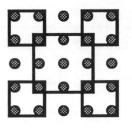

Teacher's Notes

Section G
Perimeter

Getting Started

Use your geoboard and geoband to mark off a one-unit length. Remind children that the raised lines on their geoboards mark off unit lengths. You may wish to show children a piece of cardboard or straw that is the same length as one unit on the geoboard. Use the unit to measure the length of a pencil, a shoe, or some line segments drawn on the chalkboard. Record the measurements on the chalkboard.

Stretch a geoband to the length of 1 unit; then to 2, 3, and 4 units. Then make several figures of different sizes on your geoboard, and help children count the number of units around each figure.

Using the Worksheets

Use the worksheets in the order shown at the right.	Counting Units	50
	Identifying One-Unit Lengths	51
	Distance Around	52
	More Difficult Shapes	53
	How Long is Each Fence?	54
	Making Fences	55
	Matching Shapes and Perimeters	56

As you progress through the worksheets, make sure children are counting the units correctly, not missing any or counting any twice. Introduce the word *perimeter,* and explain that it means the distance around a shape.

Practice

Have children make squares and rectangles on their geoboards. They first find the perimeter, and then the area. Emphasize the difference between the two kinds of measurements: perimeter is a measure of length; area is a measure of the square units that cover the inside of a shape.

Wrap-up

Ask children to measure the perimeter of figures you make on your geoboard. Then give them different perimeters and have them make the figures on their geoboards.

This is one unit long. •———•
How many units around each shape?

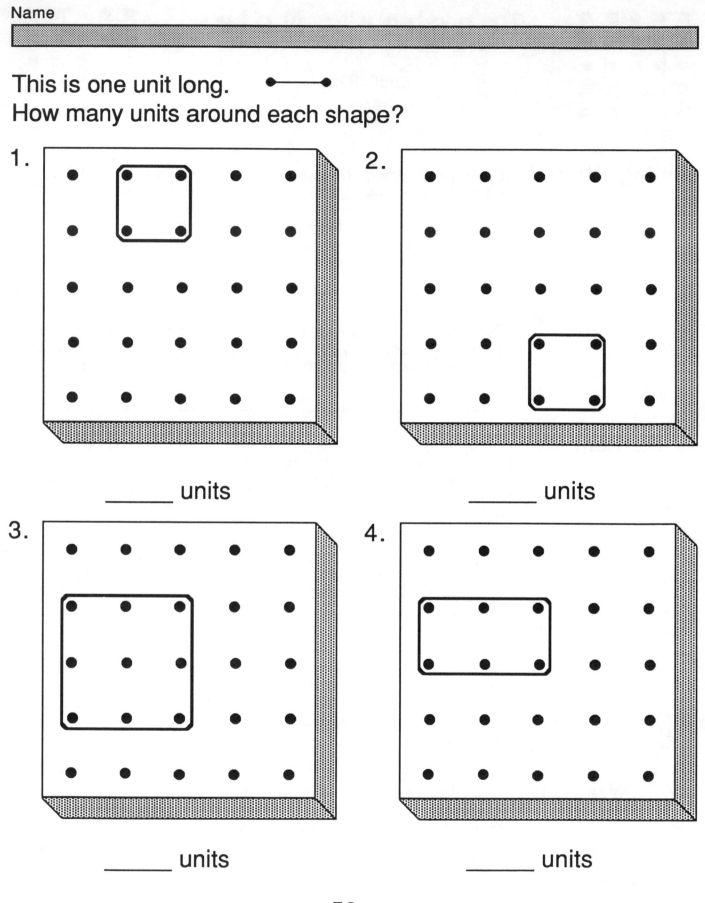

1.

_____ units

2.

_____ units

3.

_____ units

4.

_____ units

50

Name

Is each segment one unit long? Circle yes or no.

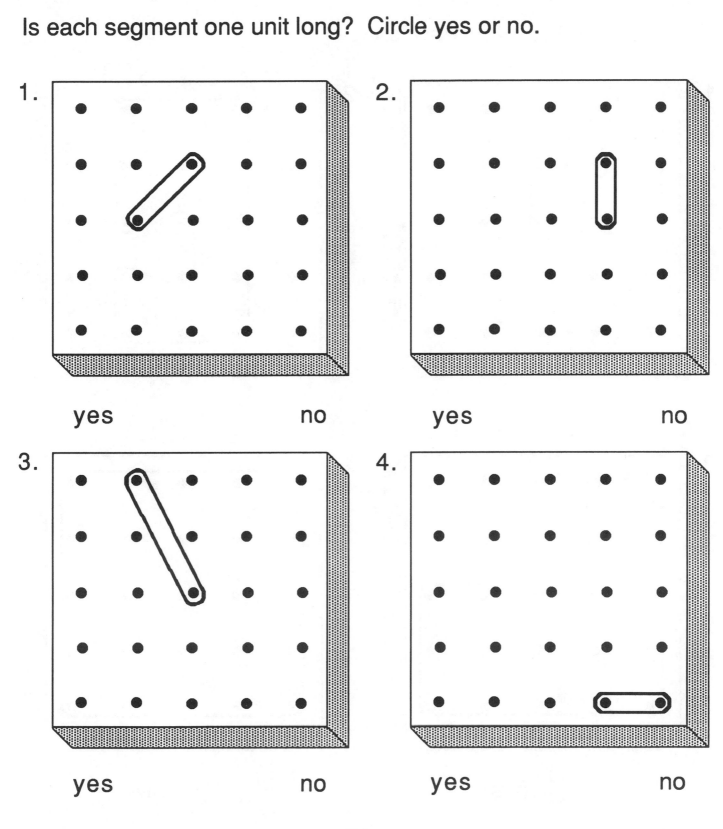

1.

yes   no

2.

yes   no

3.

yes   no

4.

yes   no

How many units around each shape?

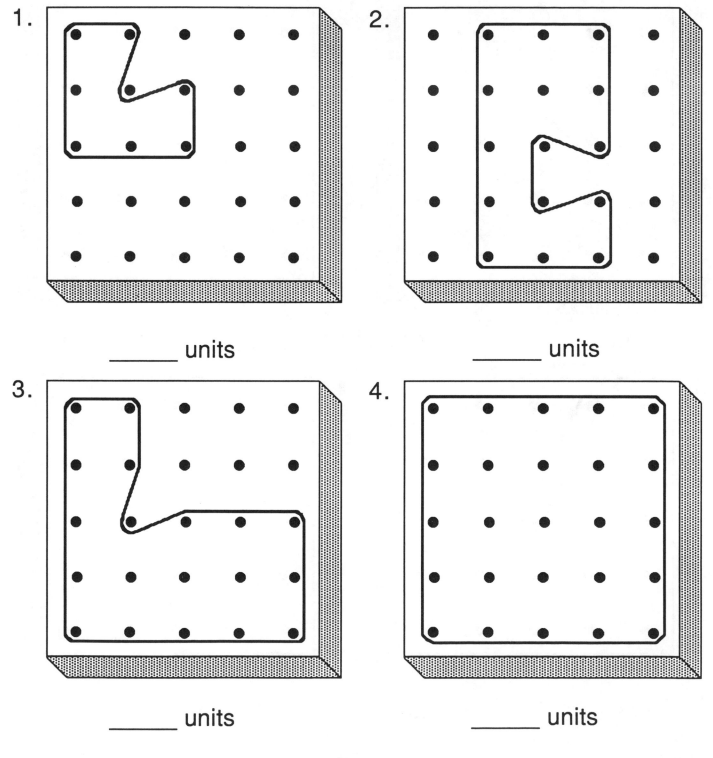

1.

_____ units

2.

_____ units

3.

_____ units

4.

_____ units

Match each shape with the number of units around it.

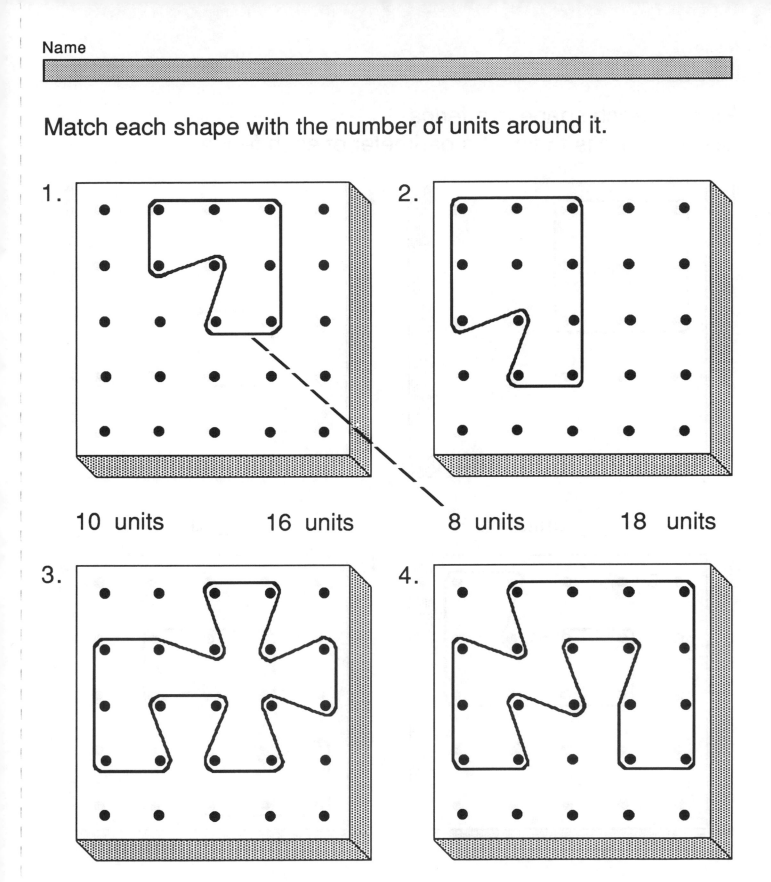

1.

2.

10 units   16 units   8 units   18 units

3.

4.

Pretend each shape is a fence.
Count the units to find the perimeter of each fence.

1.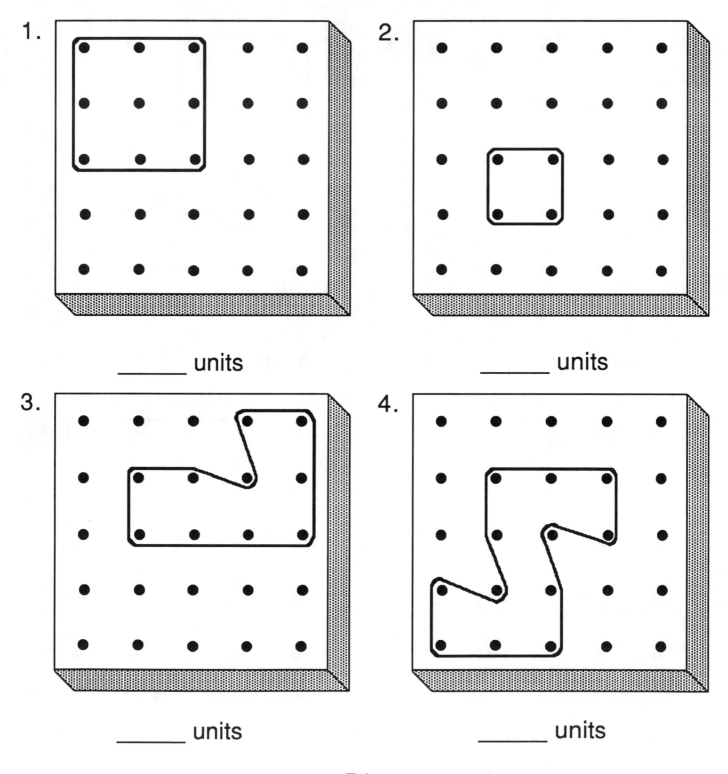

_____ units

2.

_____ units

3.

_____ units

4.

_____ units

Make each fence.

1. Use 4 units.

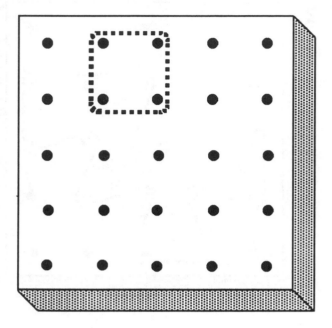

2. Use 8 units.

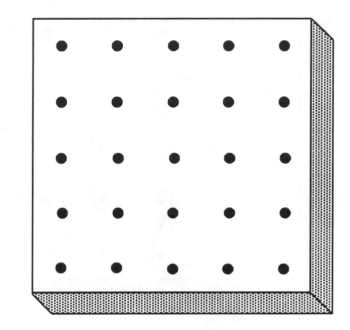

3. Use 10 units.

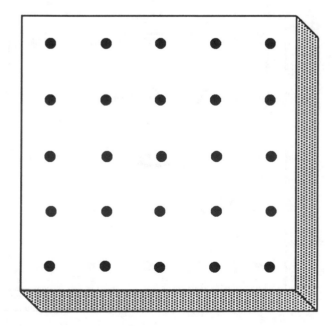

4. Use 16 units.

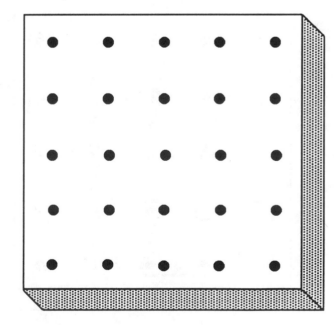

Shade the inside of each fence a different color.

Match each shape with its perimeter.

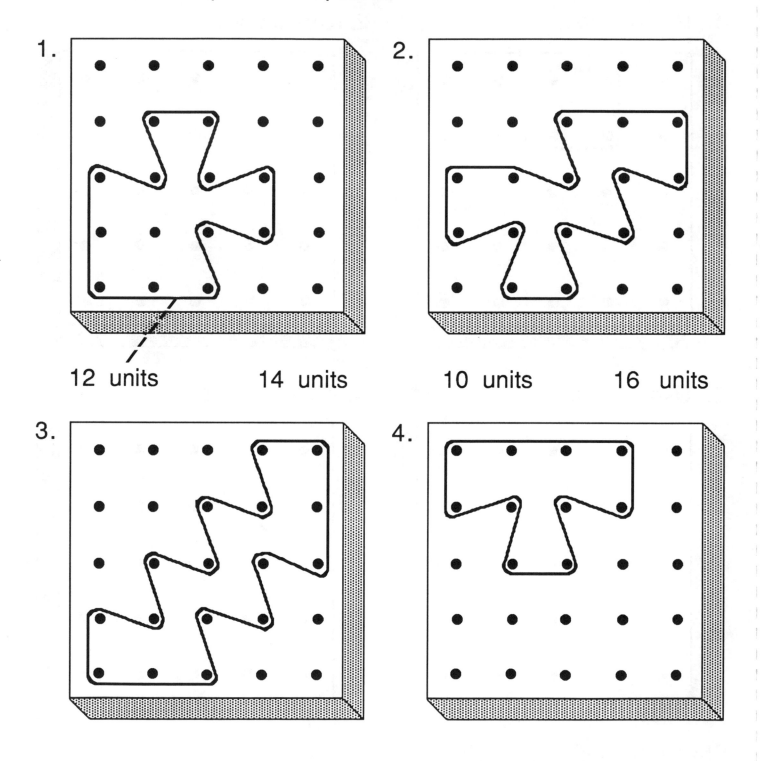

1.

12 units   14 units

2.

10 units   16 units

3.

4.

1.

2.

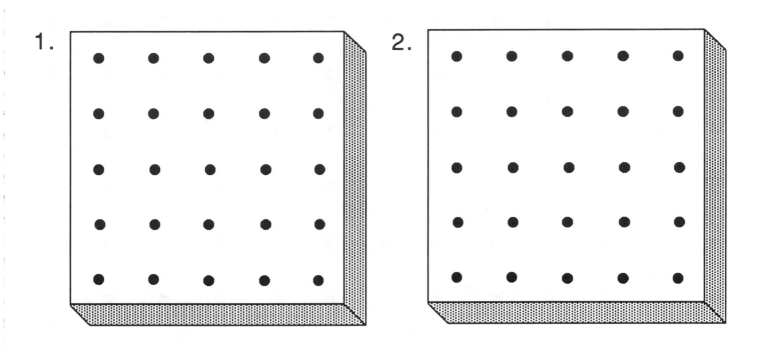

3.

4.

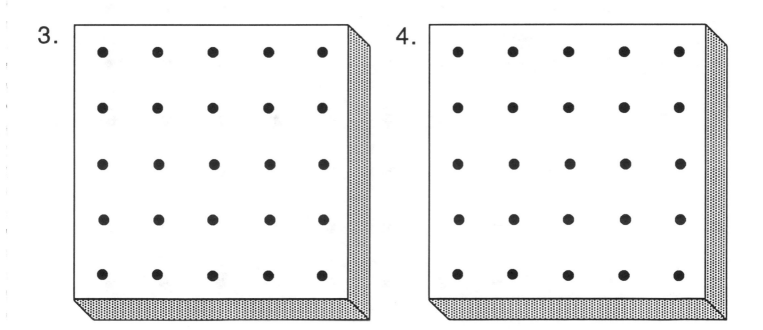

Name